Area & Perimeter
A complete workbook with lessons and problems

By Maria Miller

Contents

Preface

Hello! I am Maria Miller, the author of this math book. I love math, and I also love teaching. I hope that I can help you to love math also!

I was born in Finland, where I also grew up and received all of my education, including a Master's degree in mathematics. After I left Finland, I started tutoring some home-schooled children in mathematics. That was what sparked me to start writing math books in 2002, and I have kept on going ever since.

In my spare time, I enjoy swimming, bicycling, playing the piano, reading, and helping out with Inspire4.com website. You can learn more about me and about my other books at the website MathMammoth.com.

This book, along with all of my books, focuses on the conceptual side of math... also called the "why" of math. It is a part of a series of workbooks that covers all math concepts and topics for grades 1-7. Each book contains both instruction and exercises, so is actually better termed *worktext* (a textbook and workbook combined).

My lower level books (approximately grades 1-5) explain a lot of mental math strategies, which help build number sense — proven in studies to predict a student's further success in algebra.

All of the books employ visual models and exercises based on visual models, which, again, help you comprehend the "why" of math. The "how" of math, or procedures and algorithms, are not forgotten either. In these books, you will find plenty of varying exercises which will help you look at the ideas of math from several different angles.

I hope you will enjoy learning math with me!

Introduction

Area & Perimeter Workbook includes lessons and exercises suitable for third grade.

Students find perimeters of polygons, including finding the perimeter when the side lengths are given, and finding an unknown side length when the perimeter is given. They learn about area, and how to measure it in either square inches, square feet, square centimeters, square meters, or just square units if no unit of length is specified.

Students also relate area to the operations of multiplication and addition. They learn to find the area of a rectangle by multiplying the side lengths, and to find the area of rectilinear figures by dividing them into rectangles and adding the areas.

We also study the distributive property "in disguise." This means using an area model to represent $a \times (b + c)$ as being equal to $a \times b$ plus $a \times c$. The expression $a \times (b + c)$ is the area of a rectangle with side lengths a and $(b + c)$, which is equal to the areas of two rectangles, one with sides a and b, and the other with sides a and c.

Multiplying by Whole Tens is a lesson about multiplication such as 3×40 or 90×7. It is put here so that students can then use their multiplication skills to calculate areas of bigger rectangles.

Then we solve many area and perimeter problems. That is necessary so that students learn to distinguish between these two concepts. They also get to see rectangles with the same perimeter and different areas or with the same area and different perimeters.

I wish you success with teaching math!

Maria Miller

Geometry Resources on the Internet

I encourage you to use some of these free resources that can make geometry so much fun!

Free Worksheets for Area and Perimeter
Create customizable worksheets for the area and the perimeter of rectangles. Options include using images, generating word problems, or problems where the student writes an expression for the area using the distributive property.
http://www.homeschoolmath.net/worksheets/area_perimeter_rectangles.php

FunBrain: Shape Surveyor Geometry Game
An easy game that practices finding either the perimeter or area of rectangles.
https://www.funbrain.com/games/shape-surveyor

Perimeter Shapes Shoot Game
"Shoot" the shapes that have the given perimeter.
http://www.sheppardsoftware.com/mathgames/geometry/shapeshoot/PerimeterShapesShoot.htm

Perimeter at Gordon's
Work out the perimeter of the shapes. There are many options to choose from.
http://www.wldps.com/gordons/Perimeter.swf

Shape Explorer
Find the perimeter and area of odd shapes on a rectangular grid.
http://www.shodor.org/interactivate/activities/ShapeExplorer/

Area of Rectangle
Drag the corners of the rectangle and see how the side lengths and areas change.
http://illuminations.nctm.org/ActivityDetail.aspx?ID=46

Build a Robot
Collect six parts to build your own robot by answering questions about perimeter.
http://www.learnalberta.ca/content/me3us/flash/lessonLauncher.html?lesson=lessons/12/m3_12_00_x.swf

Area Shapes Shoot Game
Click on the shapes that show the given area.
http://www.sheppardsoftware.com/mathgames/geometry/shapeshoot/AreaShapesShoot.htm

Math Playground: Party Designer
You need to design areas for the party, such as a crafts table, food table, seesaw, and so on, so they have the given perimeters and areas.
https://www.mathplayground.com/PartyDesigner/index.html

Zoo Designer
You have been hired to design five enclosures for the animals at a local zoo. Use your knowledge of area and perimeter to design the correct enclosures and to earn your ZooDesigner Points.
http://mrnussbaum.com/zoo/

Area Blocks
Cover your grid with shapes before your opponent does.
http://www.mathplayground.com/area_blocks.html

Area and Perimeter Builder

Create your own rectangular shapes using colorful blocks and explore the relationship between perimeter and area. You can choose to show the side lengths to understand how a perimeter works. You can also use two work areas (grids) to compare the area and perimeter of two shapes side-by-side. Lastly, challenge yourself in the game screen to build shapes or find the area of various figures.
http://phet.colorado.edu/sims/html/area-builder/latest/area-builder_en.html

Math Playground: Measuring the Area and Perimeter of Rectangles

Amy and her brother, Ben, explain how to find the area and perimeter of rectangles and show you how changing the perimeter of a rectangle affects its area. After the lesson, you will use an interactive ruler to measure the length and width of 10 rectangles, and to calculate the perimeter and area of each.
http://www.mathplayground.com/area_perimeter.html

XP Math: Find Perimeters of Parallelograms

This online quiz shows you parallelograms and rectangles, and you need to calculate the perimeter, including typing in the right unit, and not using the altitude of the parallelogram.
http://www.xpmath.com/forums/arcade.php?do=play&gameid=10

Area and the Distributive Property Quiz

Use area models to represent the distributive property in finding area of rectangles.
https://www.khanacademy.org/math/cc-third-grade-math/cc-third-grade-measurement/cc-third-grade-area-distributive-property/e/area-and-the-distributive-property

Perimeter

Perimeter means the "walk-around measure," or the distance you go if you walk all the way around the figure.

The word comes from the Greek word *perimetros*. In it, *peri* means 'around' and *metros* means 'measure'.

To find the perimeter of this rectangle, <u>count the units</u> as you go around the figure. You can think of running or hopping around the figure.

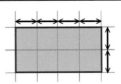

The units are marked with little arrows in the picture. The top side is four units long. The right side is two units long. Make sure you understand that!

So, what is the perimeter? _____ units

Here it is trickier to count those little units. Be careful!

How many units is the perimeter? _____ units

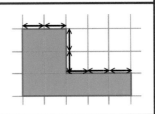

1. Find the perimeter of these figures. Your answer will be so many units. P means perimeter.

a. 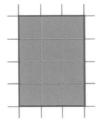 P = ___*units*___	b. 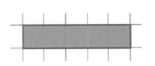 P = _____	c. 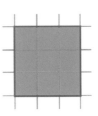 P = _____
d. 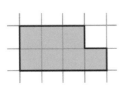 P = _____	e. 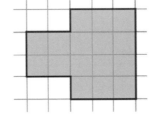 P = _____	f. 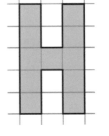 P = _____

2. Measure with a ruler to find the perimeter of these figures in centimeters.

a.

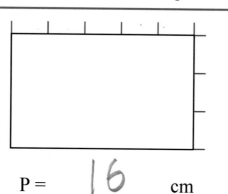

P = _____16_____ cm

b.

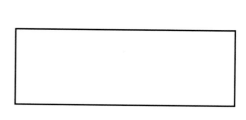

P = _____ cm

c.

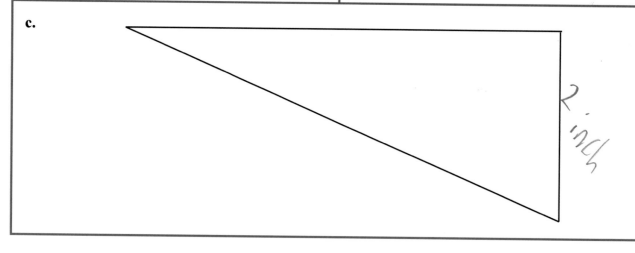

3. Measure with a ruler to find the perimeter of these figures in inches.

a.

P = _____ in.

b.

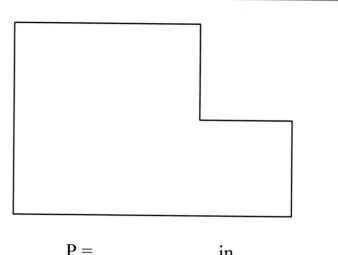

P = _____ in.

You can trace the ruler below and tape it on an existing ruler or cardboard!
Or cut it out after you have finished the neighboring page.

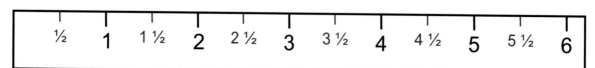

To find the perimeter, simply **add all the side lengths.** How many units is the perimeter of the triangle on the right? It is 8 + 9 + 10 units, or _____ units.	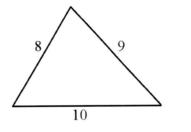
Often you need to figure out some side lengths that are not given. What side lengths are not given? The perimeter is _____ cm.	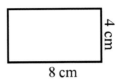

Don't forget the <u>unit of measurement</u> in your answer.

If the side lengths are in centimeters, the perimeter will be so-many *centimeters.*

If the side lengths are "plain numbers" without any particular unit, then the perimeter is so-many *units.*

4. Find the perimeter. Notice: some side lengths are not given! Don't forget to use either "cm" or "in." or "units" in your answer.

a. 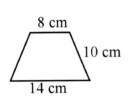 P = _____ *u n i t s*	b. 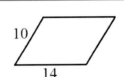 P = _____	c.  P = _____
d. P = _____	e. 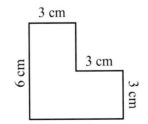 P = _____	f. 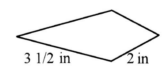 P = _____

5. Find the perimeter...

   a. ...of a square with 7-in. sides

   b. ...of a square with 13-cm sides

Problems with Perimeter

The perimeter of a rectangle is 30 cm. Its one side is 9 cm. How long is the other side?

We can write a "how many more" addition, or an addition with an unknown:

$9 + \underline{?} + 9 + \underline{?} = 30$

You could guess and check to solve it. But, there is also an easier way. Just think: the two sides, 9 and $\underline{?}$, form **half** of the perimeter. So, $9 + \underline{?} = 15$.

Thinking either way, we can solve that $\underline{?} = 6$ cm.

?

9 cm   9 cm

?

1. Solve. Write an addition with an unknown for each problem.

a. The perimeter of this rectangle is 20 cm. Its one side is 6 cm. How long is the other side?

6 cm

?

_____

Solution: $\underline{?}$ = _____

b. The perimeter of this rectangle is 44 cm. Its one side is 15 cm. How long is the other side?

?

15 cm

_____

Solution: $\underline{?}$ = _____

c. The one side of this rectangle is 12 in. Its perimeter is 82 in. How long is the other side?

?

12 in.

_____

Solution: $\underline{?}$ = _____

d. The perimeter of this square is 12 in. How long is its side?

?

?

_____

Solution: $\underline{?}$ = _____

14

2. Solve.

a. The perimeter of this square is 44 cm.
How long is the side of the square?

?

b. Find the perimeter of this square with
12-inch sides.

c. Find the perimeter of this L-shape. Notice that some
side lengths are not given.

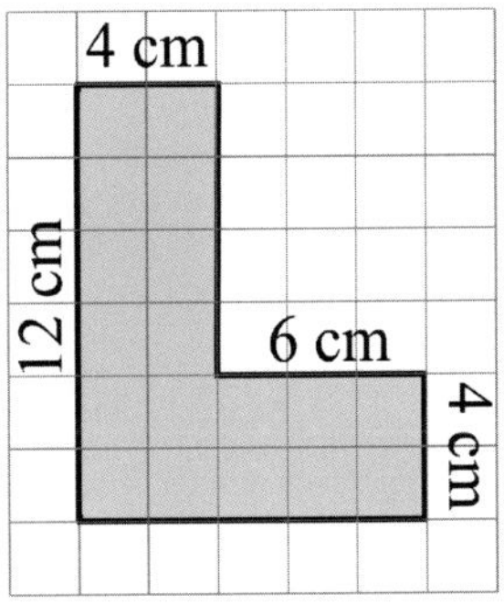

4 cm

12 cm

6 cm

4 cm

3. The parking lot of a school is in the shape shown here.
Each little square in the image has a side of 10 feet.
What is the perimeter of the parking lot?

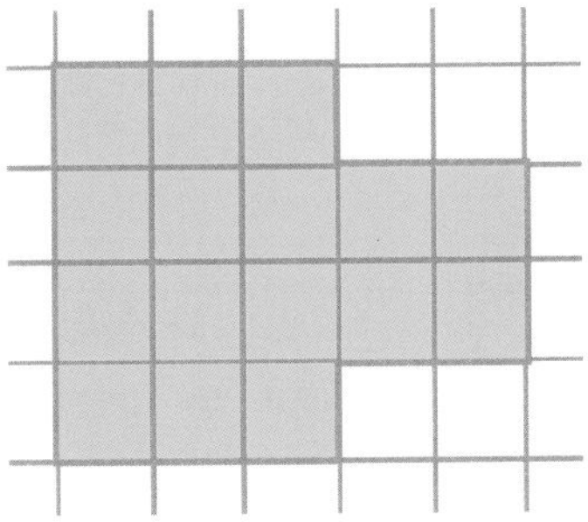

4. Kyle's house measures 25 feet wide and 35 feet long.
What is its perimeter?

5. Mandy wants a rectangular garden with a perimeter of 18 meters.
One side of the garden is 3 m.
How long should the other side be?

6. Draw many different rectangles that all have a perimeter of 24 units. Then, write the side lengths of those rectangles in the table.

Hint: The two sides of the rectangle form half of the perimeter, which is 12 units.

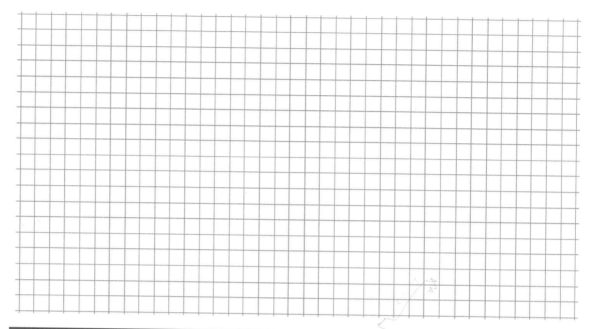

One side	Other side	Perimeter
3 units	9 units	24 units
		24 units
		24 units
		24 units

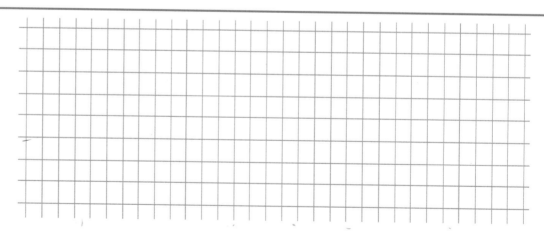

Draw a shape here that is **not** a rectangle, and that has a perimeter of
a. 8 units   b. 10 units   c. 14 units

Getting Started with Area

How many little squares do you need to cover this rectangle?

That is its **area**. Area has to do with covering, and it is measured in little squares, which we call *square units*.

The area of this rectangle is _28_ square units.

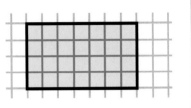

1. How many square units is the area of these figures?

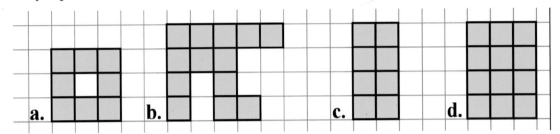

a. The area is _8_ square units.

c. The area is _9_ square units.

b. The area is _13_ square units.

d. The area is _12_ square units.

You can use multiplication to find the area of a rectangle. Notice how there are rows and columns of squares!

There are 3 rows, and 8 columns. We multiply $3 \times 8 = 24$.

The area of this rectangle is 24 square units.

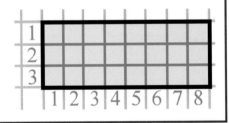

2. Write a multiplication to find the area. "A" means area.

a.

$2 \times 5 = 10$

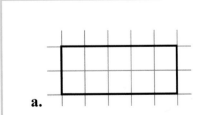

A = _10_ square units.

b.

$3 \times 3 = 9$

A = _9_ square units.

c.

$3 \times 6 = 18$

A = _18_ square units.

3. Find the areas of these figures.

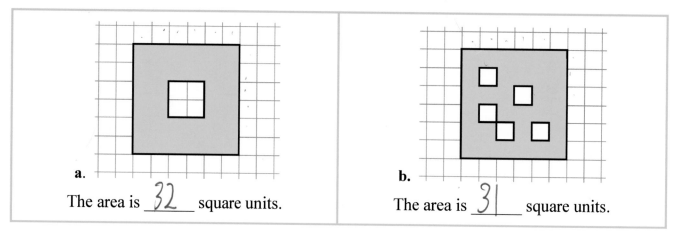

a. The area is _15_ square units.

c. The area is _10_ square units.

b. The area is _12_ square units.

d. The area is _17_ square units.

4. Find the areas.

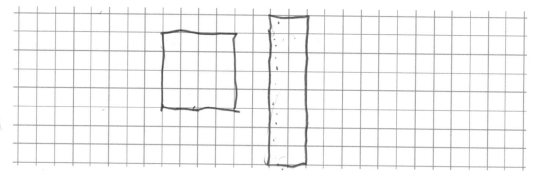

a.
The area is _32_ square units.

b.
The area is _31_ square units.

5. Draw two rectangles or squares with an area of 16 square units.

6. Draw two rectangles with an area of 24 square units.

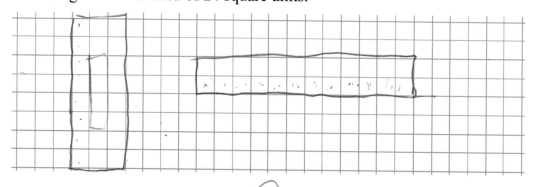

More about Area

To find the area of this figure, we can divide the shape into two rectangles. We then use two multiplications, and add their results.

$3 \times 2 + 3 \times 5 = 6 + 15 = 21$ square units

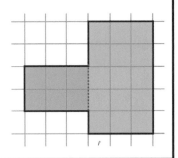

Here, can you think how to use multiplication and *subtraction* to find the shaded area? Do not look at the answer (below) yet! Think first!

It is $4 \times 5 - 2 \times 2 = 20 - 4 = 16$ square units.

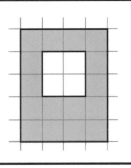

1. Write two multiplications to find the total area.

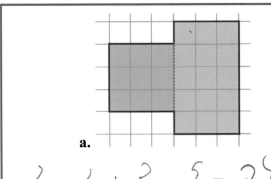

a.

$3 \times 3 + 3 \times 5 = 24 \,_u^2$

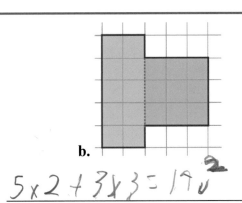

b.

$5 \times 2 + 3 \times 3 = 19 \,_u^2$

c.

$3 \times 2 + 3 \times 5 = \boxed{21,}$

missing

d.

$4 \times 5 = 70 \; 72 \times 1 = 28$

The total area of this rectangle is $3 \times 8 = 24$ square units. But notice: we can write the longer side of the rectangle as a sum $(3 + 5)$. Then, its area would be written as $3 \times (3 + 5)$.

But if we think of it as two rectangles, we can write the area as $3 \times 3 + 3 \times 5$.

So, thinking of it as a one rectangle or two rectangles, we get:

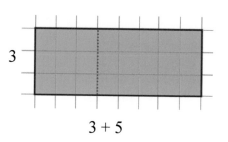

3

3 + 5

$$3 \times (3 + 5) \quad = \quad 3 \times 3 \quad + \quad 3 \times 5$$

area of the   area of the   area of the
whole rectangle   first part   second part

2. Write a number sentence for the total area, thinking of one rectangle or two.

a.

___ × (___ + ___) = ___ × ___ + ___ × ___

area of the whole rectangle   area of the first part   area of the second part

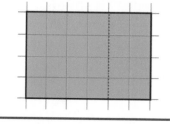

b.

___ × (___ + ___) = ___ × ___ + ___ × ___

area of the whole rectangle   area of the first part   area of the second part

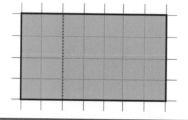

c.

___ × (___ + ___) = ___ × ___ + ___ × ___

area of the whole rectangle   area of the first part   area of the second part

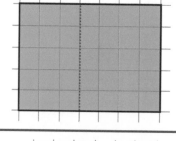

d.

___ × (___ + ___) = ___ × ___ + ___ × ___

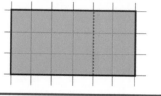

e.

___ × (___ + ___) = ___ × ___ + ___ × ___

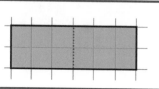

3. Now it is your turn to draw the rectangle. Fill in.

a.

$$3 \times (2 + 4) \quad = \quad \underline{\quad} \times \underline{\quad} \quad + \quad \underline{\quad} \times \underline{\quad}$$

area of the   area of the   area of the
whole rectangle   first part   second part

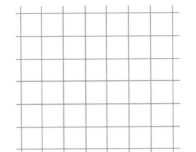

b.

$$5 \times (1 + 4) \quad = \quad \underline{\quad} \times \underline{\quad} \quad + \quad \underline{\quad} \times \underline{\quad}$$

area of the   area of the   area of the
whole rectangle   first part   second part

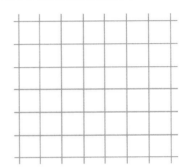

c.

$$4 \times (3 + 1) \quad = \quad \underline{\quad} \times \underline{\quad} \quad + \quad \underline{\quad} \times \underline{\quad}$$

area of the   area of the   area of the
whole rectangle   first part   second part

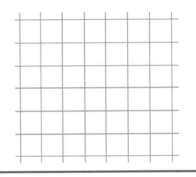

d.

$$\underline{\quad} \times (\underline{\quad} + \underline{\quad}) \quad = \quad 3 \times 2 \quad + \quad 3 \times 1$$

area of the   area of the   area of the
whole rectangle   first part   second part

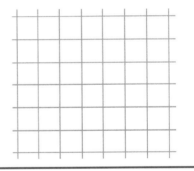

e.

$$\underline{\quad} \times (\underline{\quad} + \underline{\quad}) \quad = \quad 2 \times 5 \quad + \quad 2 \times 2$$

area of the   area of the   area of the
whole rectangle   first part   second part

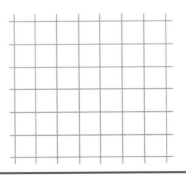

4. Find the areas of the figures.

a. Find the shaded area. Write a number sentence for the area.

_____

_____

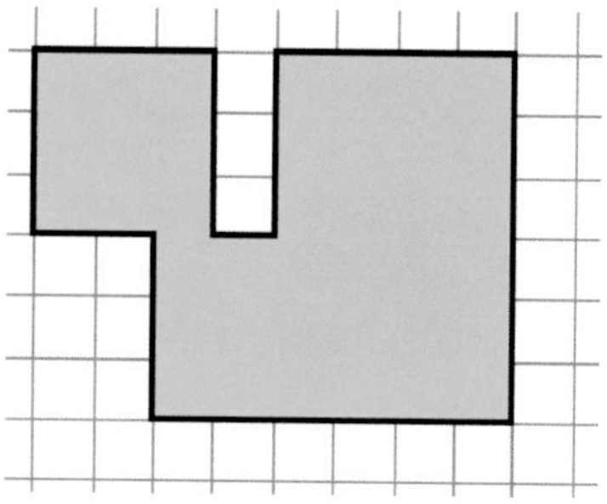

b. Find the shaded area.
   Think what operations you can use this time.
   Write a number sentence for the area.

_____

_____

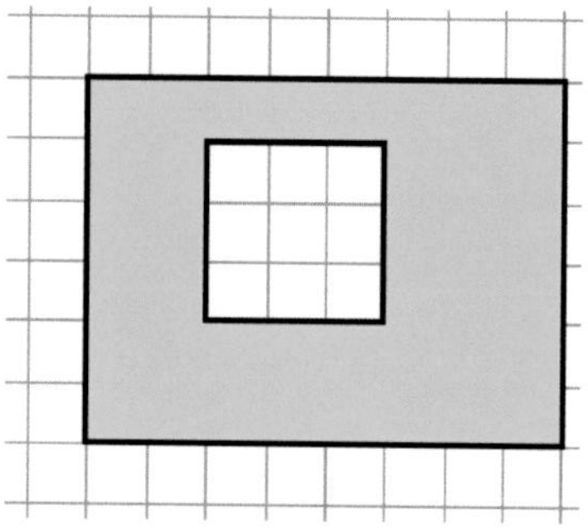

c. Find the shaded area *(not including the school)*. Write a number sentence for the area.

_____

_____

_____

yard

school

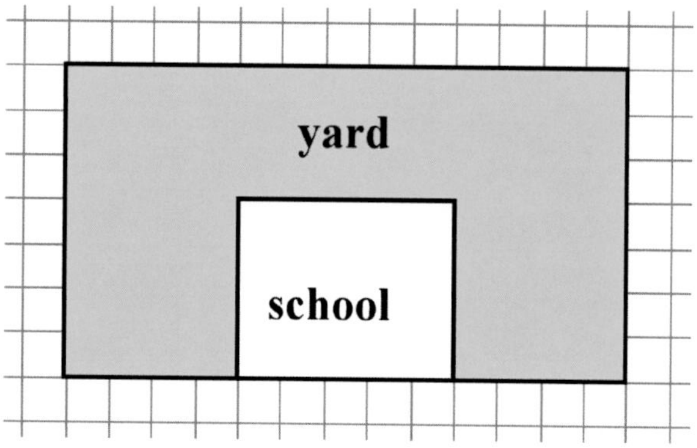

Puzzle Corner   The area of this shape is 32 squares. Your task is to write a number sentence for the area.

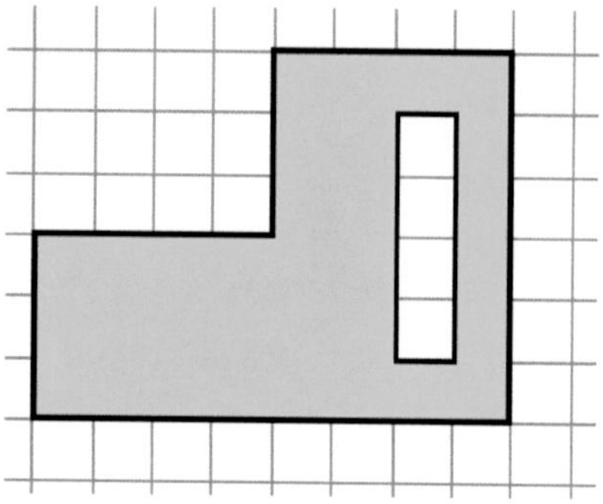

Multiplying by Whole Tens

1. Fill in the missing parts of the multiplication table of 10. Think of counting by tens!

9 × 10 = _____	14 × 10 = _____	19 × 10 = _____
10 × 10 = _____	15 × 10 = _____	20 × 10 = _____
11 × 10 = _____	16 × 10 = _____	21 × 10 = _____
12 × 10 = _____	17 × 10 = _____	22 × 10 = _____
13 × 10 = _____	18 × 10 = _____	23 × 10 = _____

There is a pattern: *Every answer ends in* _____ . Also, there is something special about the number you multiply times 10, and the answer. Can you see that?

SHORTCUT

To multiply any number by *ten*, write the number, and tag one zero on it.

For example: <u>78</u> × 10 = <u>78</u>**0**   or   10 × <u>49</u> = <u>49</u>**0**

---

2. Multiply.

a. 10 × 11 = _____	**b.** 10 × 99 = _____	**c.** 82 × 10 = _____
56 × 10 = _____	18 × 10 = _____	10 × 0 = _____

Note: If the number you multiply by 10 ends in zero, you still need to tag one zero on the answer.

For example: <u>30</u> × 10 = <u>30</u>**0**

---

3. Multiply.

a. 10 × 5 = _____	**b.** 10 × 90 = _____	**c.** 17 × 10 = _____
10 × 50 = _____	100 × 9 = _____	17 × 1 = _____

This rectangle illustrates the multiplication 7 × 20.
It has 7 rows and 20 columns.

We could COUNT the little squares to find its area.
Or, we could solve 7 × 20 by adding 20 repeatedly.

But here is yet a different way to think about it:
Let's divide this big rectangle into TWO
smaller rectangles that each are the size 7 × 10.

Each of the two rectangles has an area of 7 × 10 = 70.
So, in total their area is 70 + 70 = 140.

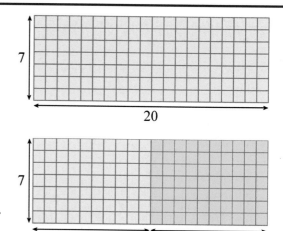

4. Solve.

a. Solve 8 × 20 by dividing this rectangle into TWO equal parts.

Parts: _____ × _____ and _____ × _____ . The total area is _____ .

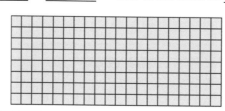

b. Solve 5 × 30 by dividing this rectangle into THREE equal parts.

Parts: _____ × _____ and _____ × _____ and _____ × _____ . The total area is _____ .

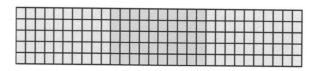

c. Solve 7 × 30 by dividing this rectangle into THREE equal parts.

Parts: _____ × _____ and _____ × _____ and _____ × _____ . The total area is _____ .

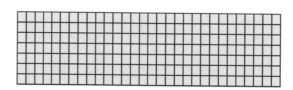

d. Solve 4 × 40 by dividing this rectangle into parts.

Parts: _____ . The total area is _____ .

24

We can solve multiplication problems, such as 5 × 60, by repeated addition.

$$5 \times 60 = 60 + 60 + 60 + 60 + 60$$

(60 added five times)

5. Solve these multiplications by repeated addition. But also look for a pattern and a shortcut. Can you find it?

a. 3 × 40 = _____	**b.** 2 × 80 = _____	**c.** 4 × 40 = _____
d. 5 × 30 = _____	**e.** 5 × 70 = _____	**f.** 3 × 80 = _____

Here's another idea for solving multiplication problems, such as 5 × 60.

Notice: 60 is equal to 6 × 10, isn't it?
So, to solve 5 × 60, we can multiply 5 × 6 × 10.

And 5 × 6 × 10 is the same as 30 × 10.
Then, 30 × 10 is just 30 with a zero tagged on the end of it... or 300.

6. Break each multiplication into another where you multiply three numbers, one of them being 10. Multiply and fill in.

a. 7 × 90 = _7_ × _9_ × 10 = _6 3_ × 10 = _____	**b.** 4 × 80 = ___ × ___ × 10 = ___ × 10 = _____
c. 6 × 40 = ___ × ___ × 10 = ___ × 10 = _____	**d.** 9 × 90 = ___ × ___ × 10 = ___ × 10 = _____
e. 30 × 6 = 10 × ___ × ___ = 10 × ___ = _____	**f.** 80 × 3 = 10 × ___ × ___ = 10 × ___ = _____

Study the **shortcut** for multiplying by whole tens.	
Example 1. 6×20 Multiply $\boxed{6 \times 2} = 12$. Tag a zero to 12, to get 120.	**Example 2.** 90×7 Multiply $\boxed{9 \times 7} = 63$. Tag a zero to 63, to get 630.

7. Multiply using the shortcut.

a. $7 \times 70 =$ _____	**b.** $6 \times 80 =$ _____	**c.** $40 \times 7 =$ _____
d. $50 \times 4 =$ _____	**e.** $70 \times 3 =$ _____	**f.** $3 \times 90 =$ _____

8. This rectangle is 7 units high and 80 units long. What is its area?

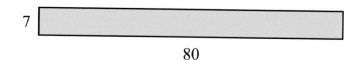

9. This rectangle is divided into 8 equal parts. What is the area of each small part?

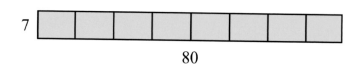

10. Find the total area of this rectangle, and also the area of each little part.

11. Find the total area.

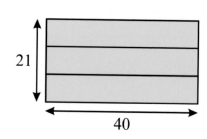

Figure out a way or two ways to solve 5×16 *without* counting all the squares.

Puzzle Corner

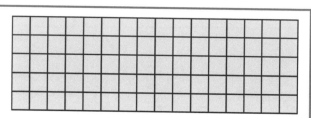

26

Area Units and Problems

Area is always measured in *squares of some size*. To find the area of a shape, we check how many squares are needed to cover the shape.

 Each side of this square measures 1 centimeter. It is a special square. It is called **a square centimeter**. We can use it to measure areas of other shapes.

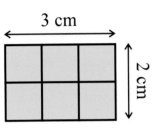 We need 6 square centimeters to cover this rectangle. So, its area is just that: 6 square centimeters. We abbreviate this as 6 cm². The elevated ^{2} indicates the "squaring."

We can also use *multiplication* to find the area:

$$3 \text{ cm} \times 2 \text{ cm} = 6 \text{ cm}^2$$

1. Write a multiplication for the area of each rectangle. Measure the sides of the rectangles in centimeters using a ruler. Do not forget the units (cm and cm²)!

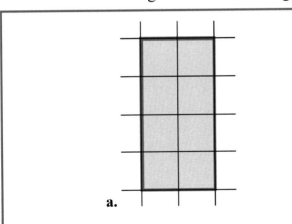

a.

A = _____ cm × _____ cm = _____ cm²

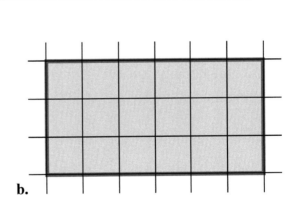

b.

A = _____ cm × _____ cm = _____ cm²

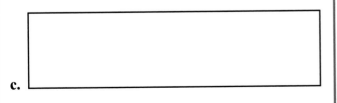

c.

A = _____

d.

A = _____

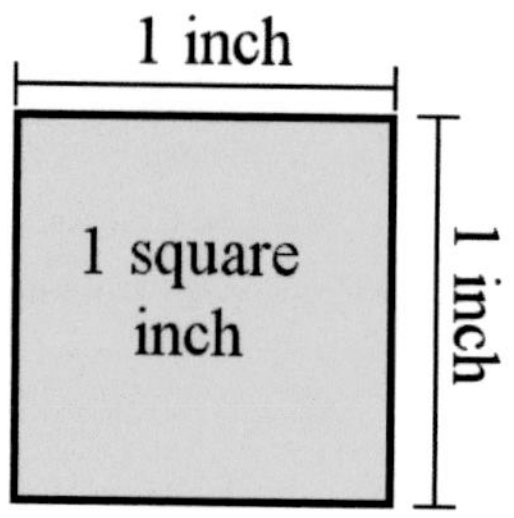

1 inch	Each side of this square measures 1 inch. It is also a special square. It is called **<u>one square inch</u>**, abbreviated as 1 sq. in. or 1 in^2. We can use it to measure areas of other shapes.

(The square shown: labeled "1 square inch" with "1 inch" across the top and "1 inch" along the side.)

2. Find the area of each rectangle. Measure in inches using a ruler. Do not forget the unit for the area.

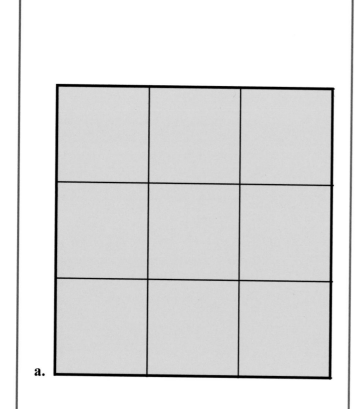

a.

A = _____ in. × _____ in. = _____ in^2

b.

A = _____ in. × _____ in. = _____ in^2

c. A = _____

The following pictures are *not* to scale. They show some other square units for area.

1 foot

This is one square foot or 1 ft².

1 m

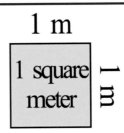

This is one square meter, or 1 m².

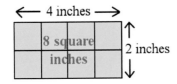

We need 8 square inches to cover this rectangle. So, its area is 8 square inches. We abbreviate this as 8 sq. in. or 8 in².

Again, use *multiplication* to find the area:

4 inches × 2 inches = 8 square inches

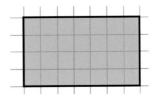

If no particular unit of length is given for the sides of a rectangle, we just use the word "unit."

The sides are 7 and 4 units, and the area is 28 *square units*.

3. Find the areas of the rectangles. Be very careful about the unit you need to use, whether square centimeters (cm²), square meters (m²), square inches (in²), or square feet (ft²).

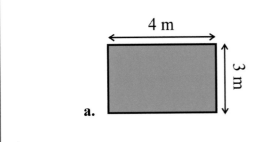

a.

A = _____

5 ft

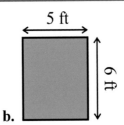

b.

A = _____

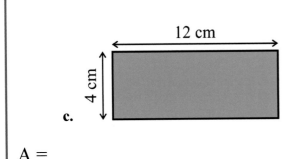

c.

A = _____

d.   8 in.

A = _____

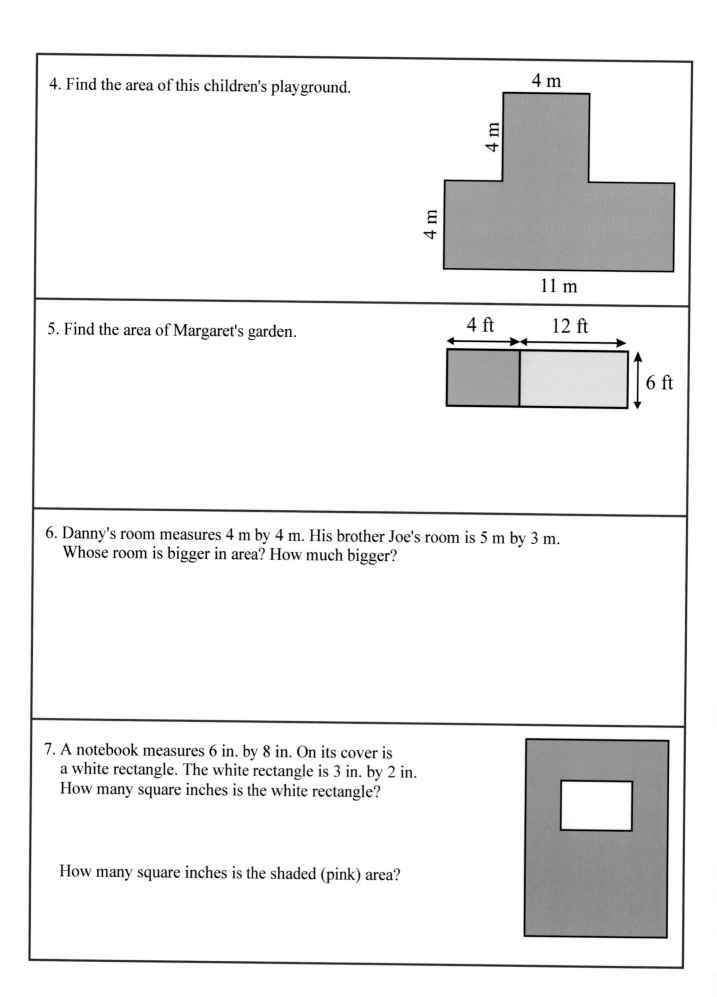

4. Find the area of this children's playground.

4 m

4 m

4 m

11 m

5. Find the area of Margaret's garden.

4 ft   12 ft

6 ft

6. Danny's room measures 4 m by 4 m. His brother Joe's room is 5 m by 3 m.
   Whose room is bigger in area? How much bigger?

7. A notebook measures 6 in. by 8 in. On its cover is
   a white rectangle. The white rectangle is 3 in. by 2 in.
   How many square inches is the white rectangle?

   How many square inches is the shaded (pink) area?

Area and Perimeter Problems

Sometimes it is easy to confuse perimeter and area. • AREA has to do with <u>covering the shape with squares</u>. Your answer will be in square centimeters, square inches, square feet, square meters, or just square units. • PERIMETER has to do with "going all the way around." Your answer will be in some unit of length, such as centimeters, meters, inches, or feet.	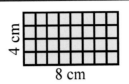 **Area:** $4 \text{ cm} \times 8 \text{ cm} = 32 \text{ cm}^2$. **Perimeter:** $4 \text{ cm} + 8 \text{ cm} + 4 \text{ cm} + 8 \text{ cm} = 24 \text{ cm}$

1. Find the area and perimeter of the rectangles.

a. 5 m, 2 m

Perimeter = _____

Area = _____

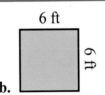

b. 6 ft, 6 ft

Perimeter = _____

Area = _____

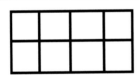

c. 4 in. wide, 2 in. tall

Perimeter = _____

Area = _____

d. A square with 3 cm sides

Perimeter = _____

Area = _____

2. Find the area and perimeter of this shape. Notice that one side length is not given. You need to figure that out.

Area

Perimeter

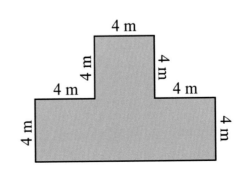

3. Find the area and perimeter of this shape.
   Notice that one side length is not given.
   You need to figure that out.

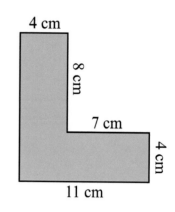

   Area

   Perimeter

4. This is a two-part lawn.

   a. Find the areas of the two parts.

   _____ and _____

   b. Find the total area.

   c. Find the perimeter.

5. Find the total area of this rectangle,
   and also the area of each little part.

   Area of each part:

   Total area:

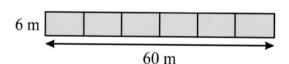

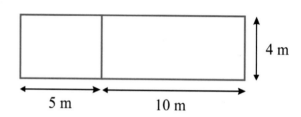

Puzzle Corner   Can you draw these rectangles? Guess and check!

   a. Draw a rectangle with an
   area of 39 squares, and
   a perimeter of 32 units.

   b. Draw a rectangle with an
   area of 56 squares, and
   a perimeter of 36 units.

More Area and Perimeter Problems

1. **a.** Find the area for each part.

_____ and _____

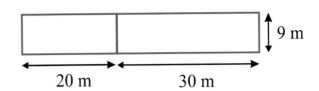

9 m

20 m   30 m

b. Find the total area.

c. Find the perimeter.

2. Make rectangles that have an area of 24 square units.

   Draw them in the grid.
   Write in the table their side lengths. One is already given.

	first side	second side	area
Rectangle 1	2 units	12 units	24 square units
Rectangle 2			24 square units
Rectangle 3			24 square units

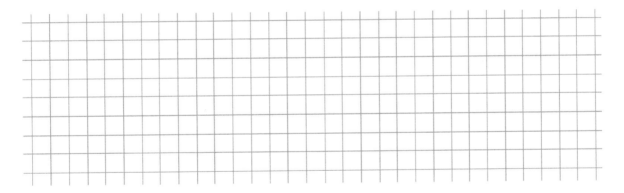

3. For each rectangle you made in #2, calculate its perimeter.

	one side	second side	area	perimeter
Rectangle 1	2 units	12 units	24 square units	units
Rectangle 2			24 square units	
Rectangle 3			24 square units	

4. Make rectangles that have a
   perimeter of 20 units.
   Hint: the two different side lengths
   add up to half of the perimeter.

   Draw them in the grid.
   Write in the table their side
   lengths. One is already given.

	first side	second side	perimeter
Rectangle 1	2 units	8 units	20 units
Rectangle 2			20 units
Rectangle 3			20 units

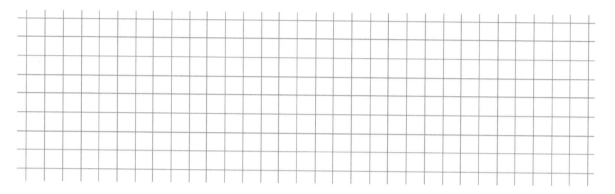

5. For each rectangle you made in #4, calculate its area.

	first side	second side	perimeter	area
Rectangle 1	2 units	8 units	20 units	square units
Rectangle 2			20 units	
Rectangle 3			20 units	

6. The image illustrates Jane's garden.

   a. Find the area of each part.

   _____ and _____

   b. Find the total area.

   c. Find the perimeter.

radishes   6 m

carrots   9 m

30 m

7. Draw and fill in.

a. Write a number sentence using the area of this two-part rectangle.

___ × (___ + ___) = ___ × ___ + ___ × ___

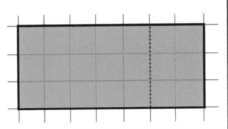

b. Draw a two-part rectangle to illustrate this number sentence.

$4 \times (3 + 5)$ = 4×3 + 4×5

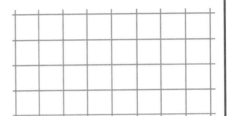

c. Fill in the missing parts, and then draw a two-part rectangle to illustrate this number sentence.

$2 \times (5 + 2)$ = ___ × ___ + ___ × ___

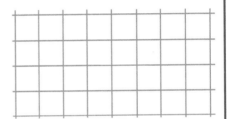

d. Fill in the missing parts, and then draw a two-part rectangle to illustrate this number sentence.

___ × (___ + ___) = 3×2 + 3×1

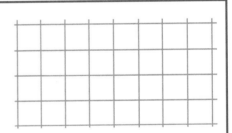

a. Write a number sentence using the area of this two-part rectangle.   Puzzle Corner

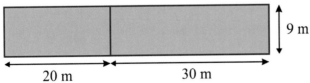

20 m   30 m

___ × (___ + ___) = ___ × ___ + ___ × ___

b. Sketch a rectangle to match $20 \times (3 + 7)$ and find its area.

35

Geometry Review

1. Fill in.

a. Write a multiplication for the area of this figure.	**b.** Draw a rectangle that has the area shown by the multiplication.
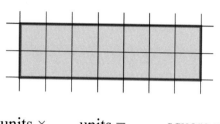	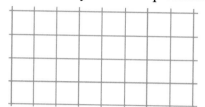
____ units × ____ units = ____ square units	4 × 5 = 20 square units

2. Find the perimeter and area of this rectangle.
   Use a centimeter ruler.

   Area:

   Perimeter:

3. Find the area and perimeter of these figures.

a. 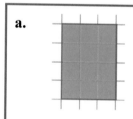 Area: Perimeter:	**b.** Area: Perimeter:

4. Write a multiplication _and_ addition for the areas of these figures.

a. 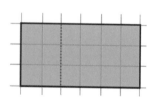 A = _____	**b.** 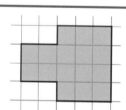 A = _____

5. Multiply using the shortcut.

a. $7 \times 70 =$ _____	**b.** $6 \times 80 =$ _____	**c.** $40 \times 7 =$ _____

6. Find the total area of this rectangle, and the area of each part.

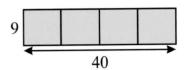

   Area of each part:

   Total area:

7. Draw and fill in.

a. Fill in the missing parts, and then draw a two-part rectangle to illustrate this number sentence.

$$3 \times (5 + 1) = \underline{} \times \underline{} + \underline{} \times \underline{}$$

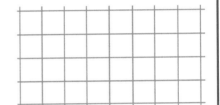

b. Fill in the missing parts, and then draw a two-part rectangle to illustrate this number sentence.

$$\underline{} \times (\underline{} + \underline{}) = 4 \times 2 + 4 \times 3$$

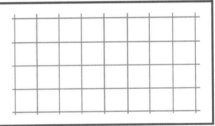

Area & Perimeter Answer Key

Perimeter, p. 11

1. a. 14 units   b. 12 units   c. 12 units
   d. 12 units   e. 18 units   f. 24 units

2. a. 16 cm   b. 16 cm   c. 12 cm + 5 cm + 13 cm = 30 cm

3. a. 6 in.   b. 10 in.

4. a. 24 units   b. 48 units   c. 3 in.
   d. 42 cm   e. 24 cm   f. 11 in.

5. a. 28 in.   b. 52 cm

To find the perimeter, simply **add all the side lengths.** How many units is the perimeter of the triangle on the right? 27 units.	
Often you need to figure out some side lengths that are not given. What side lengths are not given? The perimeter is __24__ cm.	

Problems with Perimeter, p. 14

1. a. 6 + _?_ + 6 + _?_ = 20   or   6 + _?_ = 10. The unknown _?_ = 4 cm

   b. 15 + _?_ + 15 + _?_ = 44   or   15 + _?_ = 22. The unknown _?_ = 7 cm

   c. 12 + _?_ + 12 + _?_ = 82   or   12 + _?_ = 41. The unknown _?_ = 29 in.

   d. _?_ + _?_ + _?_ + _?_ = 12   or   4 × _?_ = 12. The unknown _?_ = 3 in.

2. a. _?_ + _?_ + _?_ + _?_ = 44   or   4 × _?_ = 44. The unknown _?_ = 11 cm.

   b. The perimeter is 48 in.

   c. P = 12 cm + 4 cm + 8 cm + 6 cm + 4 cm + 10 cm = 44 cm

3. Just counting the units in the picture, the perimeter is 18 units. Since each unit is 10 feet, we get 18 × 10 feet = 180 feet. Or, you can count by tens as you count the units for the perimeter.

4. 120 feet

5. 6 m

6. Answers will vary. In each rectangle, the two side lengths should add up to 12 units (half of the perimeter).

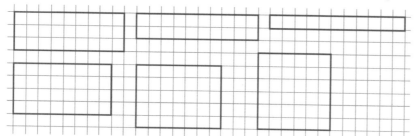

One side	Other side	Perimeter
3 units	9 units	24 units
1 unit	11 units	24 units
2 units	10 units	24 units
4 units	8 units	24 units
5 units	7 units	24 units
6 units	6 units	24 units

Puzzle corner: Answers will vary, for example:

8 units: 10 units: 14 units:

Getting Started with Area, p. 17

1. a. 8 square units   b. 13 square units   c. 8 square units   d. 12 square units

2.

a. 2 × 5 = 10 A = 10 square units.	b. 3 × 3 = 9 A = 9 square units.	c. 6 × 3 = 18 A = 18 square units.

3. a. 15 square units   b. 12 square units   c. 10 square units   d. 17 square units

4. a. 32 square units   b. 31 square units

5. The rectangles can be 1 × 16, 2 × 8, or 4 × 4.

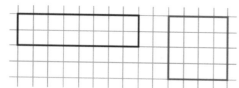

6. The rectangles can be 1 × 24, 2 × 12, 3 × 8, or 4 × 6.

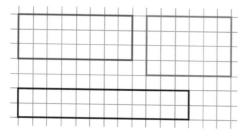

1. a. $3 \times 3 + 3 \times 5 = 24$ b. $2 \times 5 + 3 \times 3 = 19$
   c. $3 \times 5 + 2 \times 3 = 21$ d. $4 \times 5 + 2 \times 4 = 28$

2. a. $4 \times (2 + 5) = \boxed{4 \times 2} + \boxed{4 \times 5}$

   b. $4 \times (4 + 2) = \boxed{4 \times 4} + \boxed{4 \times 2}$

   c. $5 \times (3 + 4) = \boxed{5 \times 3} + \boxed{5 \times 4}$

   d. $3 \times (4 + 2) = \boxed{3 \times 4} + \boxed{3 \times 2}$

   e. $2 \times (3 + 3) = \boxed{2 \times 3} + \boxed{2 \times 3}$

3.

a. $3 \times (2 + 4) = \boxed{3 \times 2} + \boxed{3 \times 4}$ area of the whole rectangle area of the first part area of the second part	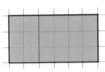
b. $5 \times (1 + 4) = \boxed{5 \times 1} + \boxed{5 \times 4}$ area of the whole rectangle area of the first part area of the second part	
c. $4 \times (3 + 1) = \boxed{4 \times 3} + \boxed{4 \times 1}$ area of the whole rectangle area of the first part area of the second part	
d. $3 \times (2 + 1) = \boxed{3 \times 2} + \boxed{3 \times 1}$ area of the whole rectangle area of the first part area of the second part	
e. $2 \times (5 + 2) = \boxed{2 \times 5} + \boxed{2 \times 2}$ area of the whole rectangle area of the first part area of the second part	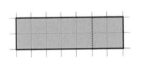

4. a. $3 \times 3 + 3 \times 6 + 3 \times 4 = 39$ square units
   b. $6 \times 8 - 3 \times 3 = 39$ square units
   c. $7 \times 4 + 5 \times 3 + 7 \times 4 = 71$ square units or $13 \times 7 - 5 \times 4 = 71$ square units

Puzzle corner. $3 \times 4 + 4 \times 6 - 4 \times 1 = 32$ squares

Multiplying by Whole Tens, p. 23

1.

$9 \times 10 = 90$	$14 \times 10 = 140$	$19 \times 10 = 190$
$10 \times 10 = 100$	$15 \times 10 = 150$	$20 \times 10 = 200$
$11 \times 10 = 110$	$16 \times 10 = 160$	$21 \times 10 = 210$
$12 \times 10 = 120$	$17 \times 10 = 170$	$22 \times 10 = 220$
$13 \times 10 = 130$	$18 \times 10 = 180$	$23 \times 10 = 230$

There is a pattern: *Every answer ends in 0*. Also, there is something special about the number you multiply times 10, and the answer. Can you see that? <u>You simply add a zero on the end of the number.</u>

2. a. 110, 560 b. 990, 180 c. 820, 0

3. a. 50, 500 b. 900, 900 c. 170, 17

4. a. Parts: 8×10 and 8×10.
   The total area is 160.
   b. Parts: 5×10 and 5×10 and 5×10.
   The total area is 150.
   c. Parts: 7×10 and 7×10 and 7×10.
   The total area is 210.
   d. Parts: 4×10 and 4×10 and 4×10 and 4×10.
   The total area is 160.

5. a. $3 \times 40 = 40 + 40 + 40 = 120$
   b. $2 \times 80 = 80 + 80 = 160$
   c. $4 \times 40 = 40 + 40 + 40 + 40 = 160$
   d. $5 \times 30 = 30 + 30 + 30 + 30 + 30 = 150$
   e. $5 \times 70 = 70 + 70 + 70 + 70 + 70 = 350$
   f. $3 \times 80 = 80 + 80 + 80 = 240$
   Multiply the numbers, then tack on the zero.

6.

a.	b.
7×90 $= \underline{7} \times \underline{9} \times 10$ $= \underline{63} \times 10 = 630$	4×80 $= 4 \times 8 \times 10$ $= 32 \times 10 = 320$
c. 6×40 $= 6 \times 4 \times 10$ $= 24 \times 10 = 240$	d. 9×90 $= 9 \times 9 \times 10$ $= 81 \times 10 = 810$
e. 30×6 $= 10 \times 3 \times 6$ $= 10 \times 18 = 180$	f. 80×3 $= 10 \times 8 \times 3$ $= 10 \times 24 = 240$

7. a. 490 b. 480 c. 280
   d. 200 e. 210 f. 270

8. The area is $7 \times 80 = 560$ square units.

9. $7 \times 10 = 70$ square units

10. The total area: $8 \times 30 = 240$ square units.
   Area of each part: $8 \times 10 = 80$ square units.

11. The rectangle is divided into thirds. Each third has the area of $7 \times 40 = 280$ square units. The total area is then $280 + 280 + 280 = 840$ square units.

Puzzle corner. Answers may vary. You can add 16 repeatedly: $16 + 16 + 16 + 16 + 16 = 80$ squares. Or, you could divide the rectangle into two parts, each having the area of $5 \times 8 = 40$. Then the total area is 80 squares.

Area Units and Problems, p. 27

1. a. $A = 2 \text{ cm} \times 4 \text{ cm} = 8 \text{ cm}^2$
   b. $A = 6 \text{ cm} \times 3 \text{ cm} = 18 \text{ cm}^2$
   c. $A = 8 \text{ cm} \times 2 \text{ cm} = 16 \text{ cm}^2$
   d. $A = 4 \text{ cm} \times 3 \text{ cm} = 12 \text{ cm}^2$

2. a. $A = 3 \text{ in.} \times 3 \text{ in.} = 9 \text{ in}^2$
   b. $A = 2 \text{ in.} \times 4 \text{ in.} = 8 \text{ in}^2$
   c. $A = 5 \text{ in.} \times 1 \text{ in.} = 5 \text{ in}^2$

3. a. $A = 4 \text{ m} \times 3 \text{ m} = 12 \text{ m}^2$
   b. $A = 5 \text{ ft} \times 6 \text{ ft} = 30 \text{ ft}^2$
   c. $A = 12 \text{ cm} \times 4 \text{ cm} = 48 \text{ cm}^2$
   d. $A = 8 \text{ in.} \times 7 \text{ in.} = 56 \text{ in}^2$

4. $A = 11 \text{ m} \times 4 \text{ m} + 4 \text{ m} \times 4 \text{ m} = 60 \text{ m}^2$

5. $A = 4 \text{ ft} \times 6 \text{ ft} + 12 \text{ ft} \times 6 \text{ ft} = 96 \text{ ft}^2$

6. Danny's room is 16 m^2. Joe's room is 15 m^2.
   Danny's room is bigger by one square meter.

7. The white rectangle has the area of $3 \text{ in.} \times 2 \text{ in.} = 6 \text{ in}^2$.
   The pink area is $6 \text{ in.} \times 8 \text{ in.} - 3 \text{ in.} \times 2 \text{ in.} = 42 \text{ in}^2$.

Area and Perimeter Problems, p. 31

1. a. perimeter 14 m; area 10 m²

   b. perimeter 24 ft; area 36 ft²

   c. perimeter 12 in.; area 8 in²

   d. perimeter 12 cm; area 9 cm²

2. a. You can divide the shape into four 4 m by 4 m squares, each having the area of 16 m². The area is then
   $16 \text{ m}^2 + 16 \text{ m}^2 + 16 \text{ m}^2 + 16 \text{ m}^2 = 64 \text{ m}^2$.
   The perimeter is 40 m.

3. For the area, divide the shape into two rectangles. That can be done in two ways.
   You could get $11 \text{ cm} \times 4 \text{ cm} + 4 \text{ cm} \times 8 \text{ cm} = 76 \text{ cm}^2$.
   or $4 \text{ cm} \times 12 \text{ cm} + 7 \text{ cm} \times 4 \text{ cm} = 76 \text{ cm}^2$.

   The perimeter is
   $4 \text{ cm} + 8 \text{ cm} + 7 \text{ cm} + 4 \text{ cm} + 11 \text{ cm} + 12 \text{ cm} = 46 \text{ cm}$.

4. a. $5 \text{ m} \times 4 \text{ m} = 20 \text{ m}^2$ and $10 \text{ m} \times 4 \text{ m} = 40 \text{ m}^2$
   b. 60 m²
   c. 38 m

5. Area of each little part is $6 \text{ m} \times 10 \text{ m} = 60 \text{ m}^2$.
   The total area is $6 \text{ m} \times 60 \text{ m} = 360 \text{ m}^2$.

Puzzle corner. a. 13 × 3 rectangle.

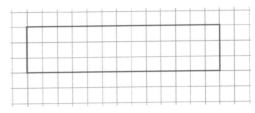

b. a 14 × 4 rectangle.

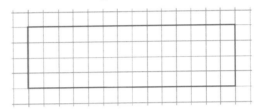

More Area and Perimeter Problems, p. 33

1. a. $20 \text{ m} \times 9 \text{ m} = 180 \text{ m}^2$ and $30 \text{ m} \times 9 \text{ m} = 270 \text{ m}^2$
   b. 450 m²
   c. 118 m

2.

	first side	second side	area
Rectangle 1	2 units	12 units	24 square units
Rectangle 2	3 units	8 units	24 square units
Rectangle 3	4 units	6 units	24 square units
Rectangle 4	1 unit	24 units	24 square units

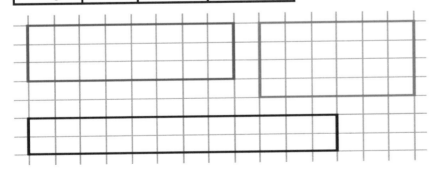

3.

	one side	second side	area	perimeter
Rectangle 1	2 units	12 units	24 square units	28 units
Rectangle 2	3 units	8 units	24 square units	22 units
Rectangle 3	4 units	6 units	24 square units	20 units
Rectangle 4	1 unit	24 units	24 square units	50 units

4.

	first side	second side	perimeter
Rectangle 1	2 units	8 units	20 units
Rectangle 2	3 units	7 units	20 units
Rectangle 3	4 units	6 units	20 units
Rectangle 4	5 units	5 units	20 units
Rectangle 5	1 unit	9 units	20 units

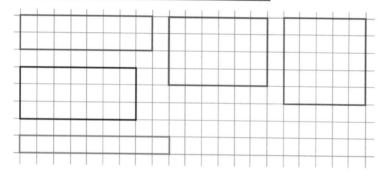

5.

	first side	second side	perimeter	area
Rectangle 1	2 units	8 units	20 units	16 square units
Rectangle 2	3 units	7 units	20 units	21 square units
Rectangle 3	4 units	6 units	20 units	24 square units
Rectangle 4	5 units	5 units	20 units	25 square units
Rectangle 5	1 unit	9 units	20 units	9 square units

6. a. $30 \text{ m} \times 9 \text{ m} = 270 \text{ m}^2$ and $30 \text{ m} \times 6 \text{ m} = 180 \text{ m}^2$.
   b. 450 m^2
   c. 90 m

More Area and Perimeter Problems, cont.

7.

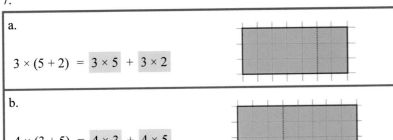

a.

$3 \times (5 + 2) = \boxed{3 \times 5} + \boxed{3 \times 2}$

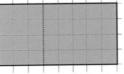

b.

$4 \times (3 + 5) = \boxed{4 \times 3} + \boxed{4 \times 5}$

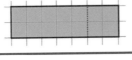

c.

$2 \times (5 + 2) = \boxed{2 \times 5} + \boxed{2 \times 2}$

d.

$3 \times (2 + 1) = \boxed{3 \times 2} + \boxed{3 \times 1}$

Puzzle corner. a. $9 \times (20 + 30) = \boxed{9 \times 20} + \boxed{9 \times 30}$

b. $10 \times 20 = 200 \text{ m}^2$

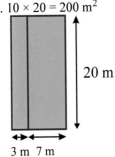

20 m

3 m 7 m

Geometry Review, p. 36

1. a. A, B, F, H, J   b. C, E, I, K, L

2. Answers will vary. Check students' answers.

3.

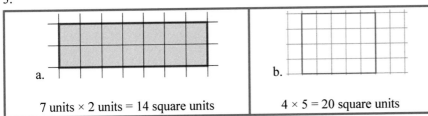

a.

7 units × 2 units = 14 square units

b.

4 × 5 = 20 square units

4. a. Area 35 cm^2 b. perimeter 24 cm

5. a. Area 12 square units; perimeter 14 units   b. Area 11 square units; perimeter 24 units

6. a. A = 3 × 2 + 3 × 4 = 18 square units   b. A = 2 × 2 + 3 × 4 = 16 square units

7. a. 490   b. 480   c. 280

8. Area of each part: 9 × 10 = 90 square units. Total area 9 × 40 = 360 square units.

9.

a.

$3 \times (5 + 1) = 3 \times 5 + 3 \times 1$

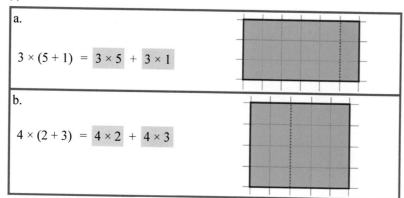

b.

$4 \times (2 + 3) = 4 \times 2 + 4 \times 3$

Appendix: Common Core Alignment

The table below lists each lesson, and next to it the relevant Common Core Standard.

Lesson	page number	Standards
Perimeter	11	3.MD.8
Problems with Perimeter	14	3.MD.8
Getting Started with Area	17	3.MD.5 3.MD.6
More about Area	19	3.MD.7
Multiplying by Whole Tens	23	3.NBT.3 3.MD.7
Area Units and Problems	27	3.MD.5 3.MD.6 3.MD.7
Area and Perimeter Problems	31	3.MD.5 3.MD.7 3.MD.8
More Area and Perimeter Problems	33	3.MD.7 3.MD.8
Geometry Review	36	3.NBT.3 3.MD.5 3.MD.6 3.MD.7 3.MD.8 3.G.1

Made in the USA
Columbia, SC
01 December 2018